Mihaela-Adriana Tita
Nicoleta Ungureanu
Ovidiu Tita

Optimizing the technology for obtaining the Cottage cheese

AF153155

Mihaela-Adriana Tita
Nicoleta Ungureanu
Ovidiu Tita

Optimizing the technology for obtaining the Cottage cheese

Cottage cheese is an unripened type of cheese, slightly acid, made from skimmed milk

LAP LAMBERT Academic Publishing

Impressum / Imprint

Bibliografische Information der Deutschen Nationalbibliothek: Die Deutsche Nationalbibliothek verzeichnet diese Publikation in der Deutschen Nationalbibliografie; detaillierte bibliografische Daten sind im Internet über http://dnb.d-nb.de abrufbar.

Alle in diesem Buch genannten Marken und Produktnamen unterliegen warenzeichen-, marken- oder patentrechtlichem Schutz bzw. sind Warenzeichen oder eingetragene Warenzeichen der jeweiligen Inhaber. Die Wiedergabe von Marken, Produktnamen, Gebrauchsnamen, Handelsnamen, Warenbezeichnungen u.s.w. in diesem Werk berechtigt auch ohne besondere Kennzeichnung nicht zu der Annahme, dass solche Namen im Sinne der Warenzeichen- und Markenschutzgesetzgebung als frei zu betrachten wären und daher von jedermann benutzt werden dürften.

Bibliographic information published by the Deutsche Nationalbibliothek: The Deutsche Nationalbibliothek lists this publication in the Deutsche Nationalbibliografie; detailed bibliographic data are available in the Internet at http://dnb.d-nb.de.

Any brand names and product names mentioned in this book are subject to trademark, brand or patent protection and are trademarks or registered trademarks of their respective holders. The use of brand names, product names, common names, trade names, product descriptions etc. even without a particular marking in this works is in no way to be construed to mean that such names may be regarded as unrestricted in respect of trademark and brand protection legislation and could thus be used by anyone.

Coverbild / Cover image: www.ingimage.com

Verlag / Publisher:
LAP LAMBERT Academic Publishing
ist ein Imprint der / is a trademark of
OmniScriptum GmbH & Co. KG
Heinrich-Böcking-Str. 6-8, 66121 Saarbrücken, Deutschland / Germany
Email: info@lap-publishing.com

Herstellung: siehe letzte Seite /
Printed at: see last page
ISBN: 978-3-659-48204-5

Zugl. / Approved by: Sibiu, University of Sibiu, 2011

OPTIMIZING THE TECHNOLOGY FOR OBTAINING THE COTTAGE CHEESE TYPE

Mihaela TIȚA[1*], Ungureanu Nicoleta[1], Ovidiu TIȚA [1]

[1]Faculty of Agricultural Sciences, Food Industry and Environment Protection, "Lucian Blaga" University of Sibiu, Dr. Ion Ratiu Street, 550024, Romania, [*]corresponding author: tita_mihaeladriana@yahoo.com,

ABSTRACT

Cottage cheese is an unripened type of cheese, slightly acid, made from skimmed milk. In order to optimize the technology for obtaining this type of cheese, we analyze the activity of lactic cultures in time, taking into consideration the pH for different samples of raw milk. The chemical composition of the milk enriched with solid substance, and the type of culture used had a great influence in the process of obtaining the Cottage cheese. Parameters that were influenced were especially the fermentation period, the pH at curd cutting and further processing. From the obtained results, the use of starter frozen cultures was the most effective option regardless of milk composition. In the full and effective optimization of technological process for obtaining Cottage cheeses, it is recommended the use of enriched milk with solid substance and inoculated with frozen starter culture.

KEYWORDS: *cheese, milk, chemical composition, fermentation*

TABLE OF CONTENT

ARGUMENT

We chose this dissertation because working in the dairy industry I consider that milk and dairy products, especially cheeses are indispensable for life because of the rich content of nutrients: protein, fat, carbohydrates, vitamins, minerals and enzymes. Milk and milk products are an important element in human nutrition, especially among children and the elderly.

The cheese is very important in human nutrition due to the nutrients contained, especially proteins, fats and minerals easily assimilated and those special sensory characteristics, taste and texture that make these dairy products one of the most consumed foods.

We have chosen Cottage cheese because it is the assortment with the largest share in the production processing and sale of cheese we manufacture.

1. Introduction

1.1. DATA FROM THE SCIENTIFIC LITERATURE ABOUT THE ISSUE UNDER STUDY

Cheese are natural products or fresh, obtained by removing the whey from the formed curd after the coagulation of whole milk, skimmed milk or partly skimmed milk, of cream or whey or from mixtures of those products.

They were, along with the milk, the main food of the various people from ancient times.

It is assumed that the first cheese was obtained by chance, the milk being stored in stomachs for long distance transport. Under the action of heat and yeasts of the stomach lining, the milk coagulated and by removing a portion of whey the cheese appeared, whose production was gradually developed.

For the ancient Greeks and Romans, cheese was a common and much appreciated food, especially the cheese from goat and sheep milk which was fatter and had a spicy taste. From Romans cheese production passed to the Gales where it developed very quickly, and then expanded on the Rhone Valley in Switzerland and from here in Germany.

Rome is considered the place of formation of several varieties of cheese which were gradually extended in many European countries (Emmental cheese would have been made around the year 58 BC by the Helvetica tribe of the Swiss Alps, as a result of the Roman's influence from the time of their invasion. By 1900, Switzerland had 700-800 cheese making places).

In our country, the production of cheese has existed for over 10,000 years, as the Thracians and Geto-dacians were animal breeders. At first it was made the sheep cheese, and then the cow cheese, the varieties depending of the influence of other people. Pressed cheese for example started being produced during the formation of the Romanian people, its technology being identical to that of Cacio- Cavallo cheese which is prepared in present Italy, but whose origin is very old. Regarding the origin of the "cheese" word, it is geto-dacian.

Classification of cheese is made according to several criteria, due to its great diversity, as a result of technical progress emerging new kinds of cheese:

➢ *Milk type:*
- Cow milk
- Sheep milk
- Buffalo milk
- Goat milk

➢ *Ripening period* (minimum duration of maturation and storage)
- Short time conservation – several days (fresh cheese)
- A week conservation (soft cheese)
- A month conservation (semi-hard and hard cheese)
- An year conservation (hard and scraped cheese)
- Pasteurized and sterilized cheese (melted cheese)

➢ *Paste consistency*
- Soft (Camambert cheese, Nasal, Alpina)
- Semi-hard (Trappist cheese, Olanda, Harghita)
- Hard (Swiss cheese, Muresana, Cedar, Parmesan)

➢ Manufacturing process:
- Fresh (fresh cottage cheese, Caraiman)
- Ripped (Zamora cheese, Vladeasa, Trappist, Olanda)
- Pickled (cow cheese, sheep, goat, buffalo)
- Scalding paste (Penteleu cheese, Dobrogea, Dalia, Rucar)
- Melted
- Kneaded cheese

Fresh cheeses are defined as products with different concentrations of fat, made from whole milk, skimmed milk or cream through acidification with lactic bacteria and/or with the clot, in which it can be added whey protein (maximum 18.5%). The product is marketed fresh, without maturation. Fat content can be established even after removing the whey.

Fresh cheeses are classified by appearance and consistency in the following categories:

- paste (quag, simple or double cream cheese)
- granular (cottage cheese);
- hard, compact, stratified (layered cheese).

COTTAGE CHEESE - technological operations

It is a type of not matured cheese, slightly acid, made from skimmed cow's milk. The main features of this type of cheese are the granular texture and flavor of sour cream (despite the low fat content) and low acidity.

Technological scheme of the Cottage cheese contains the following technological operations:

Quality reception

The milk for cheese must meet certain quality requirements laid down in national rules and standards. It must come from healthy animals, reasonably fed. Colostral milk it is not allowed in the manufacture of cheese. Also, it is not recommended the use of milk from diseased animals or treated with antibiotics because they prevent the normal development of the clotting of milk, the curd fermentation and maturation of cheese.

The milk coming from animals fed with marc, onion, wormwood etc., which taste and smell is transmitted, it is also not suitable for the manufacture of cheese. The milk for cheese must correspond in terms of numbers and type of microorganism. They have an effect on milk's acidity and some microorganisms can cause serious defects in cheese of which bloating is most common (it is caused by butyric bacteria of *Clostridium* type). The milk's acidity for cheese must not exceed 20°T; also the milk with a low acidity is not suitable.

The quality control of the milk used to make cheese is accomplished by using the following determinations:

- Sensory analysis (color, taste, smell, appearance, consistency). In the case of milk for cheese a special attention is given to taste and smell, which can be passed away to the final product;

- Physic-chemical analysis consists in the determination of density, fat, acidity and protein content;

- Microbiological analysis, which determines, in particular, the presence of coliforms and butyric bacteria.

Quantitative reception

Quantitative reception consists in weighing the milk mass at its entrance into the factory. This can be achieved through determination of the volume of the tank or by measurements taken when milk passes through a flow meter (galactomether).

Impurities removal

It performs mechanical removal of microorganisms from milk by centrifugal separation.

Churning

It is achieved by centrifugal separation using the centrifugal separator, resulting skimmed milk with a fat content of 0.1% and cream.

Skimmed milk pasteurization

Pasteurization of milk for cheese ensures the destruction of vegetative form microorganisms and the smoothing of quality product through the use of selected lactic bacteria. In addition to these advantages, pasteurization has the disadvantage that some minerals pass from the soluble state into insoluble state, thus causing a soft consistency of the curd.

According to the pasteurization temperature and hold time, there are two types of pasteurization in the usage of milk for cheese production. The first one is the low or the lasting pasteurization which is performed by heating the milk to a temperature of 63-65 ° C and its maintained so for 20-30 minutes. The second mode occurred for productivity reasons and it is the high pasteurization at 72 ° C for 15 seconds.

Milk pasteurization is done in double walled boilers with different capacities. Milk cooling is carried out, after pasteurization, at a certain temperature, depending on the cheese type and the season. The cooling is done at 32-35°C.

Given also the ambient temperature, the cooling of milk in winter is performed at a temperature of 3-4°C higher than in summer.

Preparing for coagulation

Because after pasteurization the milk's natural micro flora is destroyed it is required to seed the milk with selected lactic bacteria cultures, specific to the cheese type. The proportion of yeast (culture of lactic bacteria) added to the milk varies depending on the quality of the milk, the type cheese, the activity of the lactic cultures, season and it ranges between 0.05 and 10%.

After the introduction of yeast and its homogenization in the whole mass, the milk is left at rest so the lactic cultures can develop and also to maturate. Furthermore, after the pasteurization it is added an aqueous solution of calcium chloride at a concentration of 40% (100 ml milk 100). Lately, because of its bitter taste, the manufacturers give up more and more using it or they are replacing it with another calcium salt (phosphates).

By completing these steps, the milk is ready for clotting. This occurs under the action of coagulating enzymes of plant, animal or fungal origin or of organic acids (lactic, acetic, citric) or mineral.

9

Coagulation

This is the most important phase of the preparation of cheese. In practice is achieved by joint action of lactic acidification and enzymatic hydrolysis of coagulating enzymes.

In recent years, with the strong growth in consumption of cheese and their production, the animal coagulating enzymes that give the best results do not cover anymore the production's needs and manufacturers went over to the extraction of coagulating enzymes of microbial origin (fungal) obtained from different kinds of molds.

The animal clot has the ability to coagulate the milk with low acidity in the presence of calcium ions and to give a curd from which the whey is removed easily, thereby eliminating more lactose which is dissolved in the whey.

Under the action of clot, casein K, a component of milk casein complex which serves as its protective colloid is degraded. This way the other components of the casein complex (α and β-casein) will form a casein with calcium ions, thus obtaining the milk curd. All coagulating enzymes come in a liquid state (solutions) or in powder form. They are characterized by their coagulation strength (enzyme strength).

Processing the curd

Processing the curd aims to eliminate the appropriate amount of whey. Elimination is favored by the acidity of curd, by its most intense shredding and most high temperature. The clot which at first is soft it shrinks concurrent with the removal of whey, a phenomenon known as syneresis.

Before starting the curd's processing it is established whether it has the appropriate consistency. This operation is done through several methods:

10

1.   It is inserted a table spoon in the curd; with the concave side of the spoon is lifted some curd. A good curd for processing is flexible, it gives a straight rupture, with smooth walls and with clear whey disposal, of yellowish color;

2.   The index finger is inserted vertically into the curd and it is moved forward. A good curd for processing will split in the sense finger movement, while a too soft curd has no crack.

After cutting the curd into columns, they are cut to different sizes. It is usual that size of the curd's beans is named after the size of some cereal or some fruit (size of a walnut, hazelnut, a pea, maize, wheat, millet, etc.). The consistency of the curd depends on the pH (usually the resulted acid has a pH between 4.6 and 4.8). Cutting the curd at a lower pH gives a softer curd and by exceeding pH 4.9 it is obtained a hard curd. After cutting, the curd is allowed to stand for 10 up to 15 minutes, the surface hardens.

Bean wash

The curd is washed in three rounds as follows: first washing with water at 30-32°C with continuous stirring of the curd for 20 minutes; a second washing with water at 20 -22°C for 20 minutes; third washing with water at 2 - 8°C for strengthening the grain. If washing is less intense acid flavor is insufficiently removed.

The addition of pasteurized sour cream

The addition is made at 95°C, with 13-15% fat, in which was dissolved NaCl (1% towards the cheese), in order the final product to have 20% fat. Mixing curd with cream will improve the Cottage cheese flavor.

Cheese packaging

Cheese packing is made using paraffin carton boxes or plastic bags, with a net weigh of 200-500g.

Storage

It is realized at 2-8°C for up to 10 days.

1.2. HISTORY OF STARTER CULTURES USE

The microorganisms group involved in cheese biotechnology is diverse, but with a well defined role in different stages of the technological process. The cumulative metabolic activity of microorganisms from starter cultures and indigenous microbiota of the substrate is coordinated for the following two main purposes:

- Processing of raw material and final product maturation,
- Exclusively for maturation.

The need to obtain starter cultures emerged with the observation that these provide a better fermentation in the manufacture of cheese, if the whey from the previous day is added in the fresh milk. At first transfer culture is achieved by seeding the whey and thus is propagated a certain culture of microorganisms. Because the transferred culture was used to start fermentation and improve the activity of present culture, it was named the starter culture.

In the early twentieth century scientists began studying the microorganisms involved in cheese making. As expected, beside the lactic acid bacteria there were also found a large number of contaminations. Cultures were progressively purified and maintained so by inoculation in sterile environments. The most active of these cultures were selected and marketed as commercial starter cultures.

Starter cultures are defined as those cultures obtained from a pure culture stock and by passing through intermediate cultures (passages) are able to be used to produce fermented milk products. Biotechnological properties of starter culture, dimension and inoculums quality, as well as physic-chemical conditions of the crops are important in the process.

Starter cultures can be consisting only of a microorganism or several microorganisms.

Starter cultures of microorganisms are used to:

- Direct the biochemical processes through which the milk product is ensured to have a certain level of safety (including storage capacity);

- Ensure sensory qualities;

- Ensure in some cases also some nutritional qualities.

When using starter cultures in the dairy industry one should consider the following:

- to contain a certain number of viable microorganisms/g (ml) and a number of undesirable germs as small as possible;

- the primary and secondary metabolic products should not present any danger to human health;

- they must not contain and produce any antibiotics that are used for therapeutic purposes in humans;

- to have a certain specific activity: production of lactic acid, flavor substances, reduction of nitrogen;

- the existing microorganisms in culture must be declared with their entire scientific name.

The making of various milk products is conditioned, in most cases, by the amount and quality of present micro flora which by its action determines those biochemical processes resulting in the characteristic properties of the product.

First, through the fermentation of lactic acid it is ensured the necessary amount of acid for acidic dairy products. Also, as a result of the various microorganisms' action the milk products have a specific pleasant flavor.

In terms of technology, management of technological processes is ensured by using various selected crop with which the milk is seeded after pasteurization.

Preparation of production starter cultures (improperly called sourdoughs) involves repeated transplantations on milk, starting with a pure stock culture (inoculum) which is prepared by a specialized laboratory and is delivered to the factory in liquid or dried state.

Liquid pure stock cultures (inoculums). These cultures are more active, but harder to transport and can be stored at low temperatures (1 ... 2°C) within 10 days. They are delivered in bottles of 100 cm^3, closed with rubber or plastic stopper, packed in cardboard boxes.

It is presented in a liquid form, less consistent, yellow-white to light brown. In the warm season, in order to avoid over fermentation, calcium carbonate is added as a neutralizer which in combination with lactic acid sets free CO_2. This creates a slight pressure inside the bottles, printing to the pure culture (inoculum) a sparkling look.

Dried starter cultures (lyophilized). These are delivered in hermetically sealed vials, under vacuum or packaged in an atmosphere of CO_2, respectively nitrogen and may be stored at 4 ... 5°C for 12 months. In general, freeze-dried culture is reactivated to increase its vitality. Reactivation consists into the introduction of the vial contents in 200 cm^3 pasteurized and cooled milk and then submitted to a thermostatic process at the indicated temperature.

Pure stock cultures (inoculum) can be single cultures (consisting of one or more strains of the same species) and mixed cultures (consisting of different species).

From the liquid or lyophilized pure selected culture (inoculum) it can be obtained after reactivation by successive passages the following:

- Primary culture (primary leaven or mother-leaven)

- Secondary culture (secondary leaven)

- Tertiary culture (tertiary leaven), which can be used as a production starter culture (production leaven).

Primary culture It is obtained by inoculating pasteurized and cooled milk with pure culture (inoculum) received from a specialized laboratory. Type culture, the proportion of inoculation, temperature and thermostatic time vary depending on the product for which the production culture is used for. After the thermostatic process, the culture is quickly cooled and stored at 1 ... 2°C overnight.

Secondary culture It is obtained from the primary culture (the second day), but given that this is the second transplantation (passage), it is considered to be a more advanced stage of reactivation of pure culture (inoculum) and therefore the milk destined to a secondary culture will be inoculated with a smaller quantity of the primary culture and the thermostatic process is reduced. This culture is also kept at 1 ... 2°C for 1-2 hours.

Tertiary or production culture It is prepared from the secondary culture (the third day), according to the same protocol as for primary culture, but in terms of quantity this culture must meet the production needs and in terms of quality it must present the characteristics of good quality product (appearance, texture, taste, smell, etc.)

Tertiary or production starter culture is daily inoculated and is checked chemically, sensory and microbiological likewise. When using production starter culture one must take into account:

- Culture is pure (to contain only specific microorganisms);

- Culture is active (to produce the specific fermentation in normal time and ensure a certain acidity);

- To maintain in time its original attributes;

- The culture must be maintained for 5-6 hours before use at 1 ... 2°C in order to promote the accumulation of flavoring substances;

- Not to be older than 48 hours;

In connection to obtaining production starter cultures we make the following clarifications:

- in some cases, required by the production or poor quality of the culture, it is necessary to increase the requirement of intermediary passages (cultures), in the same manner as for the secondary culture in order to correct some defects. This is required primarily for yogurt culture in order to restore the symbiotic relationship between microorganisms;

- if the primary culture has good characteristics, it can be used directly in the preparation of the production starter culture (when using liquid stock pure cultures).

DVS cultures

Used since 1970. They are concentrated cell suspension (10^{11} cfu/g) of some strains defined together with cryoprotectors agents such as glycerol and lactose, rapidly cooled in liquid nitrogen and kept at a temperature of -196°C. For transportation to the factory there are used containers packed in dry ice. DVS cultures are made up of 3 or 4 defined strains, together or separately propagated.

Advantages offered by DVS cultures:

- DVS is a highly concentrated lactic culture, standardized for direct inoculation in the production tub and requires no activation or any other treatment or pretreatment for use, just storage;

- Convenient, eliminates the need for tanks, laboratories and sterile air systems;

- Improve the daily performance, every day with the same activity;

- Improving the quality of cheese.

Disadvantages:

- Requires an efficient refrigerator, costs are compensated by the labor for preparation of the cultures;

- Due to the small amount of developed acid it is required large amounts of coagulants. To increase the curd strength of tub should be brought up to 32-33°C instead of 30°C.

- DVS culture is more expensive.

2. SUBJECT OF RESEARCH

2.1. CURRENT STAGE

In the manufacture of cheese it has been lately detected a significantly trend of growth and diversification of production due to market requirements and competition, an increasingly fierce running between the cheese manufacturers.

The most significant developments in the manufacture of cheeses include:

- ensuring production with **quality raw materials**, by making a full milk cold chain from manufacturer to factory industrialization in order to keep active the antimicrobial substances present in milk and to preserve the native nutritional qualities, as much as possible of pure milk.

- ensuring a **raw milk with the same chemical composition** no matter of: animal breed, individuality and choice, age, lactation, health, milking, climatic conditions and season.

- the **use of starter cultures** which provides both the production of lactic acid and flavor substances, as well as the ones with inhibitory activity on harmful microorganisms: hydrogen peroxide, bactericides, microbiological cleaning of the milk by the use of impurities removal technique.

Continuous diversification of the varieties of cheese is the role of improvement and coordination of selected microorganisms' activity, used as starter cultures, able to work with indigenous microorganisms, much more adapted to the technological conditions.

2.2. PROPOSED OBJECTIVES

Because the chemical composition of milk is affected by various factors it is wanted to optimize the technological process for obtaining the Cottage cheese type by:

➤   use of a raw material milk with a optimum chemical composition for achieving a final product in accordance with the quality standards;

➤   improving manufacturing times through the use of microorganisms' cultures of the type: frozen, freeze-dried and of production. Following some tests on microorganisms' cultures will determine the best solution;

➤   getting a firmer curd for further processing.

2.3.   MATERIALS AND METHODS

It will follow the process for obtaining Cottage cheese type using three samples of raw milk:

➤   sample A: skim milk with no added substance;

➤   sample B: skimmed milk with added milk powder 10%;

➤   sample C: skim milk with added 5% sodium caseinate.

2.3.1. Milk powder

Table no. 1 Milk powder characteristics

QUALITY CHARACTERISTICS		
SENSORY FEATURES	Look	Fine dust, white-yellow color, uniform, without impurities
	Smell and taste	Specific, slightly sweet
	Moisture	6 %
	Dry matter	92%
	Milk fat	0.2%
	Lactose	0.25%
	AMDI index	Disc A
	pH	6.8-7.2
	Ash	4.5%

MICROBIOLOGICAL FEATURES	Listeria /25g	Absent
	coliforms	Absent
	YeastS	50/g
	Molds	50/g
	Staphylococus aureus	Absent
	Salmonella	Absent
PACKING	20 Kg paper bags lined with polyethylene.	
IDENTIFICATION, LABELING	The bags are identified by: number of lot / batch, date of manufacture / expiration date, product name, net weight.	
STORAGE, TRANSPORT	*Store on pallets in undercover spaces, dry and cold, without water and without penetrating odor.* It is transported in covered vehicles, clean, no foreign odors.	

B. Sodium caseinate

Table no.2 Sodium caseinate characteristics

QUALITY CHARACTERISTICS		
SENSORY FEATURES	Look	Fine dust, white color, uniform, without impurities
	Smell and taste	Specific, without taste
PHYSIC-CHEMICAL CHARACTERISTICS	Grit, mesh	90 mesh
	Moisture	6 %
	Dry	92%
	Milk fat	0.2%
	Lactose	0.25%
	AMDI Index	Disc A
	pH	6.8-7.2

	Ash	4.5%
MICROBIOLOGICAL FEATURES	Listeria /25g	Absent
	Coliforms	Absent
	Lees	50/g
	Moolds	50/g
	Staphylococus aureus	Absent
	Salmonella	Absent
PACKING	20 kg paper bags lined with polyethylene	
IDENTIFICATION LABELING	The bags are identified through: number of lot / batch, date of manufacture / expiry date, product name, net weight.	
***STORAGE,* TRANSPORT**	*Store on pallets in under cover spaces, dry and cold, without water and without penetrating odor.* Delivered in covered vehicles, clean, without foreign odors.	

Milk analysis will be done with the measuring device MilkoScan which determines:

➤ dry matter content;

➤ density;

➤ protein content;

➤ lactose;

➤ fat content;

➤ acidity

Fig no.1 MilkoScan device

The device realizes a precise analysis of fat, protein, lactose, density, total dry matter and dry non-fat matter in milk, cream and liquid dairy products. Simple operation up to 50 samples per hour.

It will be analyzed how starter cultures influence the time fermentation for each of these samples; the following starter cultures will be used:

➤ production lyophilized starter culture: XPL-20 manufacturer Chr. Hansen;

➤ production frozen starter culture: Fresco 1000-50 manufacturer Chr. Hansen;

➤ production starter culture obtained by the classic way.

The sowing with culture will be at $30 \pm 2°C$ for each sample.

Lyophilized culture: XPL-20

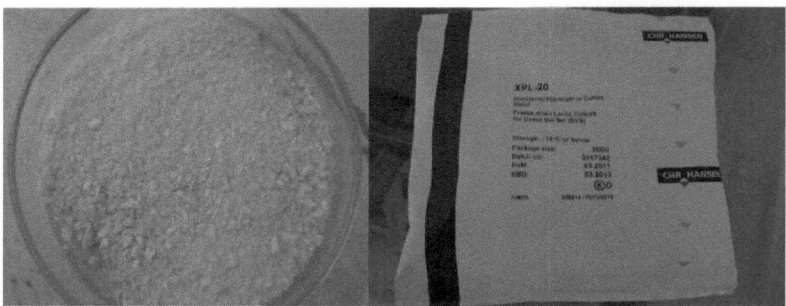

Fig. no. 2 Lyophilized culture type XPL-20

Table 3. Lyophilized culture XPL-20 characteristics

QUALITY CHARACTERISTICS		
DESCRIPTION	Culture contains	Lactococcus lactis subsp.cremoris Lactococcus lactis subsp.lactis
SENSORY FEATURES	Look	Uniform yellowish white powder
PHYSIC- CHEMICAL FEATURES	pH	>=0.9
	Temperature	30ºC
	Concentration cells	5*10 10 cfu/g
CARACTERISTICI MICROBIOLOGICE	Coliforms	<1 cfu/g
	Staph. aures	absent / 10g
	Enterococci	<20cfu/1g
	Leuconostoc	<10 cfu/1g
	Salmonella	negative in 25g
	Listeria	negativ in 25g

PACKING	*In welded under heat metallic envelopes, placed in cardboard boxes with 50 units / envelope*
IDENTIFICATION LABELING	Identification: by writing on the envelope the batch, expiration date. Packages have the following prescriptions: manufacturing company, address, product name, batch number, storage conditions, expiration date
STORAGE, TRANSPORT	*Stored at +4°C, in a clean space, without lingering odors* Transported in iceboxes, clean, without foreign odors within +4°C
USE	The product is used in the manufacture of fermented milk products

Frozen culture: Fresco 1000-1050

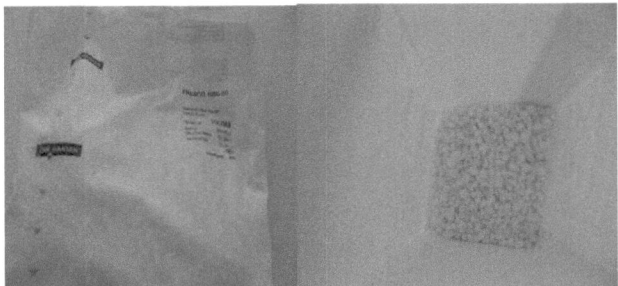

Fig. no.3 Frozen Fresco Culture 1000-1050

Table nr.4 Frozen culture Fresco 1000-1050 features

QUALITY CHARACTERISTICS		
DESCRIPTION	Culture contains	Combined mesophilic and thermophilic strains
	Culture produces	- flavor - lactic acid
SENSORY FEATURES	Look	- uniform yellowish-white granules
PHYSIC-CHEMICAL FEATURES	Temperature	30ºC
	pH	4.9-5.3
MICROBIOLOGIC AL FEATURES	Coliforms [MPN/g]	<1
	Enterococci, [cfu/g]	<10
	Lees, [cfu/g]	<1
	Molds [cfu/g]	<1
	Non lactic bacteria, [cfu/g]	<500
	Staphylococcus aureus, [cfu/g]	<1
	Salmonella	Absent in 25 g
	Listeria monocytogenes	Absent in 25 g
PACKING	*In welded under heat metallic envelopes, placed in cardboard boxes, 6 X 1000 U*	
IDENTIFICATION LABELING	Identification: by writing on the envelope the batche, expiration date Packages have the following prescriptions: producing company, address, product name, batch number, information on the composition, storage conditions,	

	expiration date.
STORAGE, TRANSPORT AND SHELF LIFE	*Frozen cultures will be stored at -45°C or below, in a clean space, without lingering odors* *Transported in iceboxes, clean, without foreign odors. At this temperature the shelf life is min. 12 months.*
USE	The culture is primarily used in cheese production.

The action of production starter cultures will be checked every 20 minutes by determining the acidity and pH of milk, respectively of the coagulated milk.

2.3.2. Determination of pH:

- determined with the pH meter Mettler Toledo1100

Fig. no.4 Mettler Toledo 1100 pH meter

2.3.3. Determination of titratable acidity

Principle of the method: the milk sample is titrated with a 0.1 N solution of sodium hydroxide in the presence of phenolphthalein as an indicator, till the sudden transfer of color to persistent pink for 30 seconds. The acidity expressed in °T is defined as the volume in ml of 0.1 N sodium hydroxide solution required to neutralize the acid in 100 ml of milk.

Needed Materials:

- Sodium hydroxide 0.1 N;
- 1% alcohol solution of phenolphthalein;
- boiled and cooled distilled water to approximately 60°C;
- 100 ml Erlenmeyer glass;
- pipette;
- burette.

Procedure

In a 100 ml Erlenmeyer glass it is introduced 10 ml of milk with the pipette. Next it is added 20 ml of warm distilled water, passed through the pipette used to measure the milk. Then it is added 3 drops of phenolphthalein. Next titrate with 0.1 N sodium hydroxide solutions under continuous stirring until it is obtained a pink color persistent for 30 seconds.

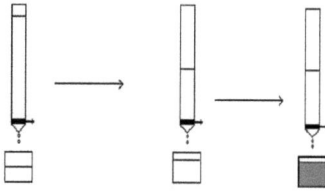

Fig.nr.5 How to determine the titratable acidity

26

Expression of results:

Acidity $° T = V \times 10$

where: V, the volume of 0.1 N NaOH used in the titration, ml;

      10 – expression factor for 100 ml milk;

- It will be used to achieve the final product the same amount of:

➢   clot;

➢   calcium chloride;

➢   salt;

➢   cream.

3. ANALYSIS FOR FINAL PRODUCT COTTAGE CHEESE

➢   Sensory analysis;

➢   Determination of acidity;

➢   Determination of salt;

➢   Determination of fat;

➢   Determination of dry matter.

3.1. SENSORY ANALYSIS

Appearance: granular, uneven.

Consistency: soft with grains of curd.

Color: white, yellowish-white

Smell and taste: pleasant, slightly salty, specific of lactic fermentation, without foreign taste and odor.

3.2. DETERMINATION OF SALT

Principle of the method: the chlorides are removed from the sample with warm water (70 . 80 ° C) and chlorine ions are titrated with a silver nitrate solution in the presence of potassium chromate as indicator.

Reagents:

- Silver nitrate solution 2.906%,; there will be weighted 2,906 g of silver nitrate which then is dissolved in about 30 cm^3. Next the solution is placed in a flask of 100 cm^3 and the content of the flask is brought to the marked volume with water and then it is homogenized.
- Potassium chromate solution 5%.

Procedure

It is weighted approximately 5 g cheese. The sample is treated with 30 cm^3 hot water (70 .. 80 ° C) till it is obtained a suspension as fine as possible. Contents is quantitatively passed through a of 100 cm^3 flask warm water (70 .. 80 ° C). Then it is cooled to 20°C and brought to the marked volume with water. Shacked well, allowed to stand for 10...15 minutes and next it is filtered through a medium porosity filter paper, in a dry glass flask of 250 cm^3.

Take with the pipette 50 cm^3 filtrate and place it in an Erlenmeyer flask of 250 cm^3. Next is added 1 cm^3 potassium chromate solution and titrate it with a silver nitrate solution 2.906% till the brick red color transfer, which doesn't disappear by shaking.

Calculation and expression of results

Sodium chloride (NaCl) = V• 100 / m • V1 [g/100 g]

where: V, the volume of the silver nitrate solution used for titration 2.906% in cm^3

V_1 product volume taken for analysis in cm^3

m - mass of the product taken for analysis, in grams.

3.3. DETERMINATION OF FAT CONTENT BY THE ACID-BUTIROMETHRIC METHOD GERBER)

Principle: the cheese sample taken for analysis it is introduced in the butiromether where it is subjected to rapid partial hydrolysis with sulfuric acid. The fat released is separated from the other components by means of centrifugation (separation is facilitated by the addition of isoamyl alcohol), and the amount expressed as a percentage is read directly from the butiromether's scale.

Needed materials:
- Butiromether for milk;
- Cellulose film soluble in sulfuric acid, with an area of 50x75 mm and 0.03. .. 0.05 mm thick;
- Automatic dispensers for sulfuric acid and 1 ml of isoamyl alcohol;
- Water bath;
- centrifuge;
- butiromethers;
- stand for butiromethers;
- sulfuric acid, density 1.817 ± 0.003;
- isoamyl alcohol, density 0.813 ± 0.002.

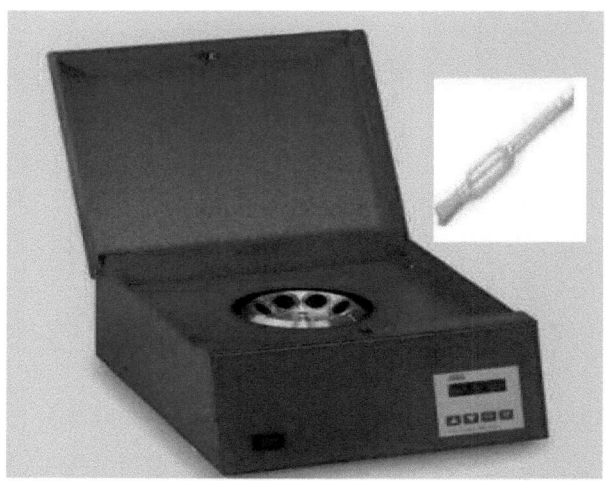

Fig. no.6 Gerber Centrifuge

Procedure

On the cellulose film, weigh 3g cheese. Insert into the butiromether 10 cm³ of sulfuric acid, in such way that it does not touch the butyromether's neck, then add in the same way hot water (30 …40 ° C) until a layer of about 6 mm it is formed above the sulfuric acid. Introduce the weighed cheese sample in the butiromether together with the folded fiber sheet. Add 1 cm³ isoamyl alcohol and hot water (30 …40 °) up to a distance of 5 mm from the lower edge of the butyromether's neck. Close the butiromether with the rubber stopper and stir until complete dissolution of the cheese. Insert the butiromether in a water bath at 65 ± 2°C and maintain for 5…10 minutes.

Insert the hot butiromether into the centrifuge. After reaching a speed of 1000-1200 rev / min (in approximately 2 minutes) centrifuge the butyromether for another 5 minutes.

Remove the butyromether from the centrifuge and place it into a water bath 65 ± 2°C for at least 3 minutes (but not more than 10 minutes).

Remove the butyromether from the water bath and read the corresponding values of the fat column upwards and downwards from the butiromether in the lowest point of the meniscus.

Calculation and expression of results

The fat content of the cheese, expressed as a percentage, is calculated using the formula:

$$\% \text{ Fat} = (B\text{-}A) \bullet 11 / m$$

In which: B - the corresponding value of the upper section of the column of fat in%

A - correspondent cut-off-value, acid fat in%

m- mass of the analyzed sample in grams.

11 - correction factor of fat content which represents the amount of product to which the milk butiromether's scale is graduated, in grams.

3.4. DETERMINATION OF DRY MATTER

Principle of the method: The evaporation of water from the sample by heating the oven at $102 \pm 2 \ ^\circ C$, until a constant value.

Materials needed:

- Electric oven

Fg. no.7 Electric oven

- Analytical balance

Fig no. 8 Analytical balance

- Water bath

Fig. no.9 Water bath

- Sea sand grain 0.15… 0.30 mm, prepared as follows: burned, washed with cold water and boiled with HCl d = 1.19 diluted 1 +1, for 30 minutes, shaking continuously. Repeat the procedure with a fresh portion of acid until it no longer gets yellow after each boiling. Sand is washed then with distilled water, until the appearance of the chloride ions. Dry at 150 ... 160°C and keep in a dark bottle.

Procedure

In a weighing vial containing a glass rod is placed about 20 g sand which then is dried in the oven at $102 \pm 2°C$ until constant weight.

After cooling the vial in the desiccator add with a pipette 10 ml of homogenized milk so that he sand can moisten evenly and weigh it. Next put it in the oven for 4-5 hours and weigh it after cooling in desiccator.

Expression of results:

Dry matter = [(m2-m) / (m1-m)] * 100%

Where: m - mass of vial with sand and rod;

m_2 – mass of vial with sand and rod, after drying;

m_1 - mass of vial with sand and rod, prior to drying.

4. RESULTS AND DISCUSSION

Determining the chemical composition of milk samples according to the MilkoScan is shown in table 5.

➤　sample A: skim milk;

➤　sample B: skimmed milk with added milk powder 10%;

➤　sample C: skim milk with added 5% sodium caseinate.

Table no. 5- MilkoScan analysis results

Determination	SAMPLE		
	A	B	C
Dry matter	8,51	9,93	10,73
Density (g/cm^3)	1,028^6	1,030	1,029
Proteins (%)	3,24	3,74	5,29
Lactose (%)	4,49	5,29	4,44
Fat (%)	0,05	0,05	0,10
Acidity (°T)	16	16	16

The influence of cultures in time according to the pH is shown in the table below for each sample of milk:

Table 6 - Skim milk sample

Time(hour)	pH analysis of Fresco frozen starter cultures for sample A	pH Analysis of freeze-dried starter culture XPL - 20 for sample A	pH analysis of classic production starter cultures for sample A
	pH	pH	pH
7.20	6,62	6,62	6,62
7.40	6,61	6,61	6,62
8.00	6,59	6,6	6,61
8.20	6,57	6,59	6,6
8.40	6,53	6,58	6,59
9.00	6,51	6,57	6,59
9.20	6,47	6,56	6,58
9.40	6,39	6,54	6,57
10.00	6,32	6,53	6,57
10.20	6,26	6,51	6,56
10.40	6,17	6,48	6,55
11.00	6,02	6.46	6,55
11.20	5,9	6,44	6,54
11.40	5,83	6,33	6,53
12.00	5,74	6,27	6,51
12.20	5,65	6,17	6,49
12.40	5,53	6,06	6,46
13.00	5,41	5,97	6,42

13.20	5,36	5,9	6,39
13.40	5,29	5,86	6,36
14.00	5,23	5,77	6,29
14.20	5,16	5,69	6,24
14.40	5,08	5,59	6,18
15.00	5,01	5,48	6,06
15.20	4,93	5,36	5,97
15.40	4,9	5,29	5,88
16.00	4,87	5,22	5,76
16.20	4,82	5,17	5,69
16.40	4,78	5,09	5,61
17.00	4,73	5,02	5,56
17.20	4,68	4,97	5,51
17.40	4,66	4,94	5,45
18.00	4,63	4,92	5,39
18.20	4,6	4,89	5,34
18.40	4,57	4,86	5,29
19.00	4,55	4,84	5,24
19.20	4,52	4,82	5,17
19.40	4,5	4,78	5,13
20.00	4,49	4,74	5,11
20.20	4,47	4,71	5.08
20.40	4,46	4,69	5,04
21.00	4,45	4,68	5,01
21.20	4,44	4,67	4,98
21.40	4,43	4,66	4,85
22.00	4,42	4,65	4,93

22.20	4,41	4,63	4,91
22.40	4,39	4,61	4,88
23.00	4,38	4,6	4,84
23.20	4,37	4,58	4,81
23.40	4,36	4,57	4,79
24.00	4,35	4,56	4,76
00.20	4,34	4,54	4,72
00.40	4,33	4,52	4,69
01.00	4,31	4,51	4,64
01.20	4,29	4,49	4,62
01.40	4,28	4,48	4,61
02.00	4,27	4,47	4,6
02.20	4,26	4,46	4,6
02.40	4,25	4,45	4,59
03.00	4,24	4,43	4,58
03.20	4,24	4,41	4,57
03.40	4,24	4,41	4,57
04.00	4,24	4,39	4,56
04.20	4,24	4,38	4,56
04.40	4,24	4,36	4,55
05.00	4,24	4,35	4,55
05.20	4,24	4,33	4,54
05.40	4,24	4,31	4,53
06.00	4,24	4,3	4,53
06.20	4,24	4,29	4,52
06.40	4,24	4,28	4,52

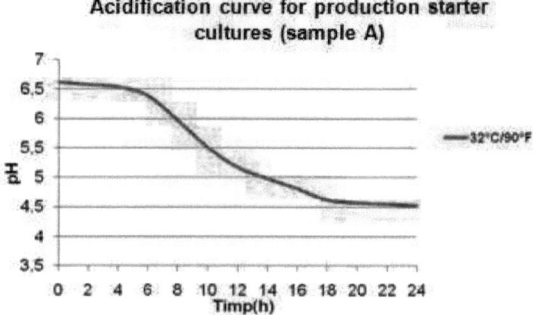

Fig. no. 10 Acidification curve of the production starter culture for sample A

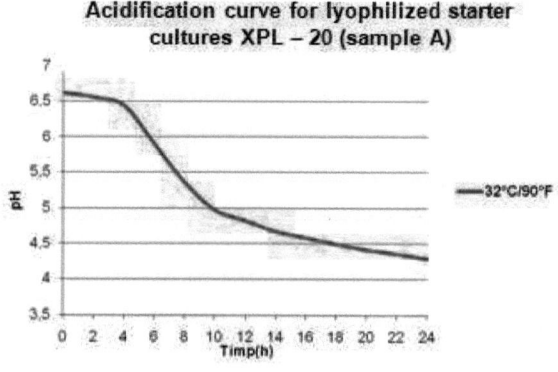

Fig. no. 11 Acidification curve for lyophilized starter cultures XPL – 20 for

sample A

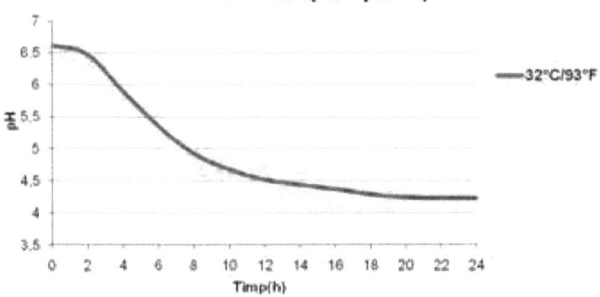

Fig. no. 12 Acidification curve of Fresco frozen cultures for sample A

Table 7 - Sample B - skimmed milk with added milk powder

Time(hour)	pH analysis of Fresco frozen starter cultures for sample B	pH analysis of freeze-dried starter culture XPL - 20 for sample B	pH analysis of classic production starter culture for sample B
	pH	pH	pH
7.20	6,63	6,63	6,63
7.40	6,61	6,63	6,63
8.00	6,58	6,62	6,63
8.20	6,54	6,61	6,62
8.40	6,51	6,61	6,62
9.00	6,48	6,6	6,61
9.20	6,45	6,59	6,61
9.40	6,36	6,4	6,6

10.00	6,31	6,41	6,6
10.20	6,19	6,43	6,59
10.40	5,89	6,45	6,59
11.00	5,78	6,44	6,58
11.20	5,7	6,46	6,58
11.40	5,59	6,39	6,56
12.00	5,47	6,31	6,54
12.20	5,34	6,29	6,52
12.40	5,12	6,19	6,5
13.00	4,96	6,07	6,47
13.20	4,83	5,81	6,45
13.40	4,81	5,66	6,42
14.00	4,7	5,44	6,39
14.20	4,69	5,14	6,36
14.40	4,67	4,96	6,32
15.00	4,58	4,94	6,3
15.20	4,55	4,92	6,27
15.40	4,51	4,87	6,23
16.00	4,49	4,84	6,18
16.20	4,48	4,8	6,12
16.40	4,47	4,79	6,06
17.00	4,46	4,74	6,02
17.20	4,45	4,71	6
17.40	4,45	4,7	5,96
18.00	4,44	4,68	5,89
18.20	4,43	4,66	5,82
18.40	4,42	4,64	5,77

19.00	4,41	4,62	5,71
19.20	4,4	4,61	5,65
19.40	4,39	4,62	5,59
20.00	4,38	4,64	5,51
20.20	4,37	4,66	5,47
20.40	4,37	4,68	5,42
21.00	4,36	4,69	5,38
21.20	4,35	4,51	5,3
21.40	4,35	4,49	5,27
22.00	4,34	4,47	5,21
22.20	4,33	4,46	5,19
22.40	4,32	4,44	5,11
23.00	4,31	4,42	5,02
23.20	4,3	4,4	4,97
23.40	4,28	4,39	4,9
24.00	4,28	4,38	4,81
00.20	4,27	4,37	4,7
00.40	4,26	4,37	4,66
01.00	4,25	4,36	4,62
01.20	4,24	4,35	4,6
01.40	4,24	4,34	4,58
02.00	4,24	4,33	4,57
02.20	4,24	4,32	4,56
02.40	4,24	4,31	4,55
03.00	4,24	4,31	4,54
03.20	4,24	4,3	4,53
03.40	4,24	4,3	4,53

04.00	4,24	4,3	4,52
04.20	4,24	4,29	4,52
04.40	4,24	4,29	4,51
05.00	4,24	4,29	4,51
05.20	4,24	4,29	4,51
05.40	4,24	4,28	4,5
06.00	4,24	4,28	4,5
06.20	4,24	4,28	4,5
06.40	4,24	4,28	4,5

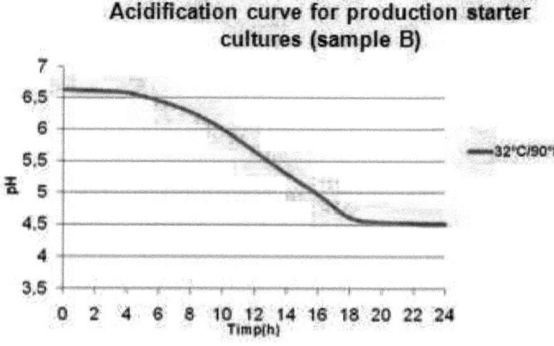

Fig. no. 13 Acidification curve of production starter culture for sample B

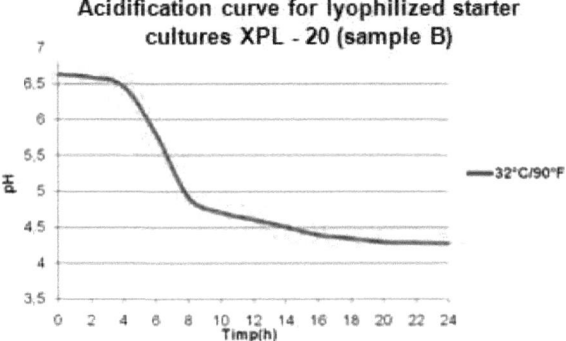

Fig. no. 14 Acidification curve of freeze-dried starter cultures XPL for sample B

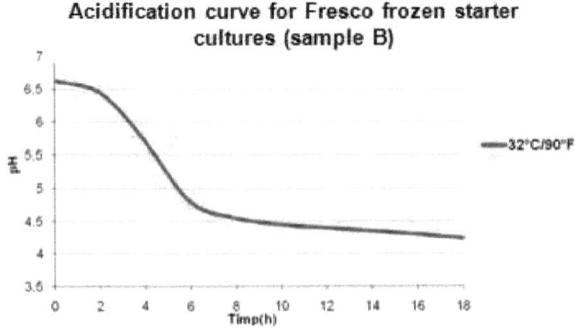

Fig. no.15 Acidification curve of Fresco frozen culture for sample B

Table 8 Sample C - skim milk with added sodium caseinate

Time hour)	pH analysis of Fresco frozen starter cultures for sample C	pH analysis of freeze-dried starter culture XPL - 20 for sample C	pH analysis of classic production starter culture sample C
	pH	pH	pH
7.20	6,74	6,74	6,74
7.40	6,72	6,73	6,73
8.00	6,67	6,71	6,7
8.20	6,63	6,69	6,71
8.40	6,59	6,68	6,69
9.00	6,54	6,66	6,68
9.20	6,41	6,65	6,67
9.40	6,3	6,63	6,65
10.00	6,14	6,6	6,64
10.20	6,95	6,57	6,62
10.40	5,75	6,52	6,61
11.00	5,55	6,45	6,59
11.20	5,24	6,3	6,57
11.40	5,19	5,97	6,54
12.00	5,07	5,86	6,52
12.20	4,97	5,72	6,5
12.40	4,89	5,56	6,47
13.00	4,82	5,4	6,45
13.20	4,74	5,31	6,43
13.40	4,72	5,23	6,41

14.00	4,7	5,09	6,39
14.20	4,65	4,97	6,37
14.40	4,6	4,88	6,35
15.00	4,56	4,79	6,32
15.20	4,54	4,69	6,29
15.40	4,51	4,66	6,23
16.00	4,49	4,64	5,96
16.20	4,46	4,62	6,21
16.40	4,44	4,6	6,16
17.00	4,39	4,58	6,12
17.20	4,35	4,55	6,07
17.40	4,34	4,54	6,02
18.00	4,34	4,52	5,97
18.20	4,33	4,49	5,92
18.40	4,33	4,48	5,87
19.00	4,32	4,47	5,81
19.20	4,31	4,46	5,76
19.40	4,31	4,45	5,7
20.00	4,3	4,43	5,65
20.20	4,29	4,43	5,6
20.40	4,29	4, 42	5,54
21.00	4,28	4,41	5,49
21.20	4,28	4,41	5,43
21.40	4,27	4,4	5,38
22.00	4,26	4,39	5,33
22.20	4,25	4,38	5,28
22.40	4,24	4,38	5,23

23.00	4,24	4,37	5,19
23.20	4,24	4,35	5,14
23.40	4,23	4,33	5,09
24.00	4,23	4,32	5,05
00.20	4,23	4,31	5,01
00.40	4,23	4,3	4,96
01.00	4,23	4,3	4,88
01.20	4,23	4,28	4,82
01.40	4,23	4,28	4,79
02.00	4,23	4,27	4,76
02.20	4,23	4,26	4,72
02.40	4,23	4,26	4,7
03.00	4,23	4,25	4,73
03.20	4,23	4,25	4,68
03.40	4,23	4,25	4,66
04.00	4,23	4,25	4,65
04.20	4,23	4,25	4,64
04.40	4,23	4,25	4,63
05.00	4,23	4,25	4,62
05.20	4,23	4,25	4,61
05.40	4,23	4,25	4,6
06.00	4,23	4,25	4,59
06.20	4,23	4,25	4,59
06.40	4,23	4,25	4,59

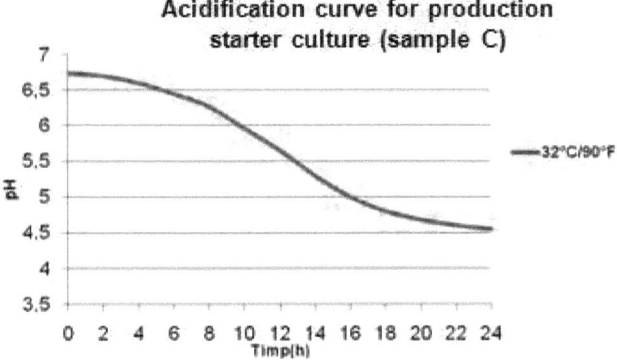

Fig. no. 16 Acidification curve of production starter culture for sample C

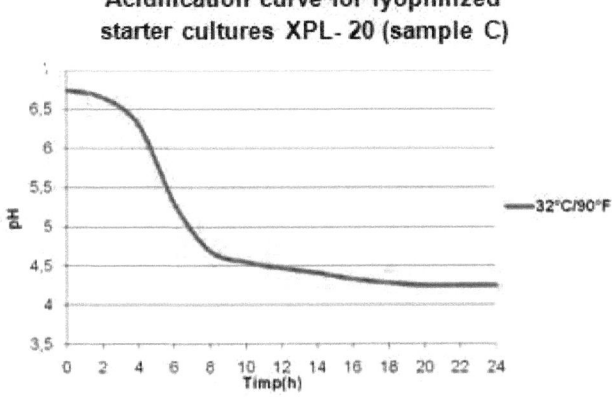

Fig. no. 17 Acidification curve of freeze-dried starter cultures XPL- 20 for

sample C

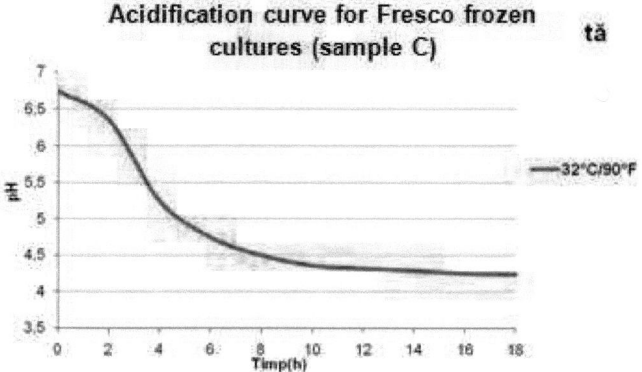

Fig. no. 18 Acidification curve of Fresco frozen culture for sample C

Appearance and consistency of the curd after cutting:

A. curd obtained with production starter culture:

Fig. no. 19 Curd beans produced with production starter culture

B. curd obtained with lyophilized culture XPL -20

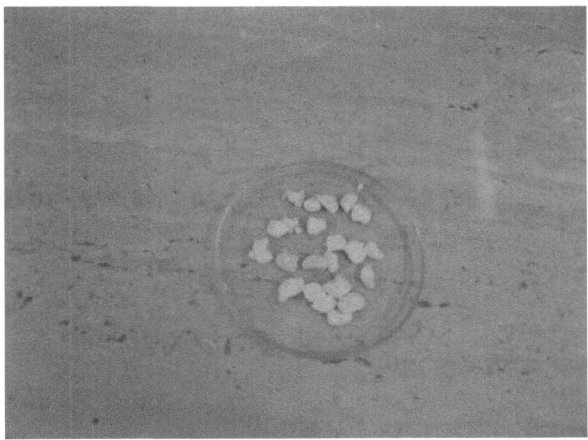

Fig. no.20 Curd made with lyophilized culture XPL-20

C. Curd obtained with frozen culture Fresco 1000-50

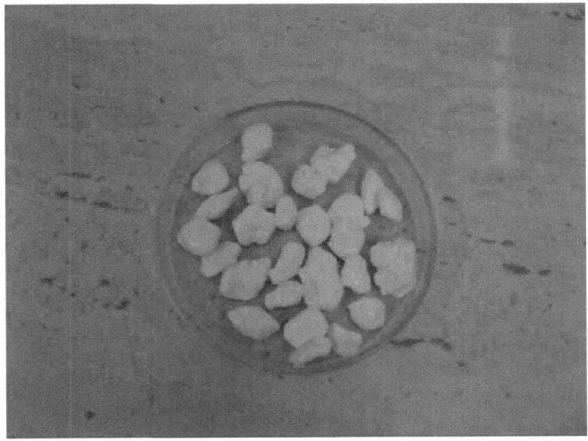

Fig. no.21 Curd made with frozen culture Fresco 1000-50

Table no. 9 Characteristics of Cottage cheese obtained by the use of different production starter cultures

Determination	COTTAGE CHEESE		
	SAMPLE - A	SAMPLE - B	SAMPLE - C
Sensory features	According to the technical specification of the product	According to the technical specification of the product	According to the technical specification of the product
Dry matter	18,24	20,34	20,08
Acidity(°T)	72	74	70
Proteins (%)	15,39	16,87	16,77
Fat(%)	4,5	4,5	4,5
Lactose (%)	2,00	2,68	2,24
Salt	1,05	0,92	0,98

5. CONCLUSIONS

Milk's chemical composition by enriching it with dry matter, but also the type of culture used had a great influence in the process of obtaining the Cottage cheese. The influenced parameters are, in particular the fermentation time, the pH of the curd at cutting and further processing.

Taking into consideration the obtained results, the use of frozen starter cultures is the most effective option regardless of the milk's composition because the fermentation is produced in less time than compared to the other types of cultures (freeze-dried and production starter cultures) for all milk samples taken for analysis.

SAMPLE - A: - production starter culture 15.2 h
- lyophilized culture 11.4 h
- frozen culture 9h

SAMPLE - B: - production starter culture 15.8 h
- lyophilized culture 9.4 h
- frozen culture 7.3 h

SAMPLE - C: - production starter culture 15.3 h
- lyophilized culture 8h
- frozen culture 6.2 h

After sowing the milk samples with frozen starter culture, the beans formed were larger, firmer, glossier compared to those formed after seeding with the other type of cultures.

In conclusion, for a full and effective optimization of the technological process of obtaining Cottage cheeses, it is recommended the use of enriched milk with dry matter and inoculated with frozen starter culture.

6. REFERENCES

Beresford, T. and A. Williams, 2004, The microbiology of cheese ripening, Cheese: chemistry, physics and microbiology, 1, 287-317.

Callanan, M. and R. Ross, 2004, Starter Cultures: Genetics, Cheese: chemistry, physics and microbiology, 1, 149-161.

Carroll, R. Cheesemaking made easy: 60 delicious varieties, Garden Way 1996

Chamba, J. F. and F. Irlinger, 2004, Secondary and adjunct cultures, Cheese: chemistry, physics and microbiology, 1, 191-206.

Delahunty, C. and M. Drake, 2004, Sensory character of cheese and its evaluation, Cheese: chemistry, physics and microbiology, 1, 455-487.

Donnelly, C., 2004, Growth and survival of microbial pathogens in cheese, Cheese: chemistry, physics and microbiology, 1, 541-559.

Fox, P. and P. McSweeney, 2004, Cheese: an overview, Cheese: chemistry, physics and microbiology, 1, 1-18.

Fox, P. F. Cheese: General aspects, Academic Press 2004

Huppertz, T., V. Upadhyay, et al., 2006, Constituents and properties of milk from different species, Brined cheeses, 1-42.

Johnson, M. E., 2001, Cheese products, FOOD SCIENCE AND TECHNOLOGY-NEW YORK-MARCEL DEKKER-, 345-384.

Le Quéré, J. L., 2004, Cheese flavour: instrumental techniques, Cheese: chemistry, physics and microbiology, 1, 489-510.

Lucey, J., 2004, Formation, structural properties and rheology of acid-coagulated milk gels, Cheese: chemistry, physics and microbiology, 1, 105-122.

McGrath, S., G. Fitzgerald, et al., 2004, Starter cultures: bacteriophage, Cheese: chemistry, physics and microbiology, 1, 163-189

McSweeney, P., G. Ottogalli, et al., 2004, Diversity of cheese varieties: an overview, Cheese: chemistry, physics and microbiology, 2, 1-23.

O'Brien, N. and T. O'Connor, 2004, Nutritional aspects of cheese, Cheese: chemistry, physics and microbiology, 1, 573-581.

Parente, E. and T. Cogan, 2004, Starter cultures: general aspects, Cheese: chemistry, physics and microbiology, 1, 123-147.